Universe

Written by John Farndon
Illustrated by Rob Jakeway

First published in 1998
This edition published in 1999 by
Parragon
Queen Street House
4, Queen Street
Bath
BA1 1HE

ISBN: 0-75253-074-7

Printed in Italy

Produced by Miles Kelly Publishing Ltd
Unit 11
Bardfield Centre
Great Bardfield
Essex
CM7 4SL

Designer: Diane Clouting
Editor: Linda Sonntag
Artwork commissioning: Branka Surla
Project manager: Margaret Berrill
Editorial assistance: Lynne French
with additional help from Jenni Cozens and Pat Crisp

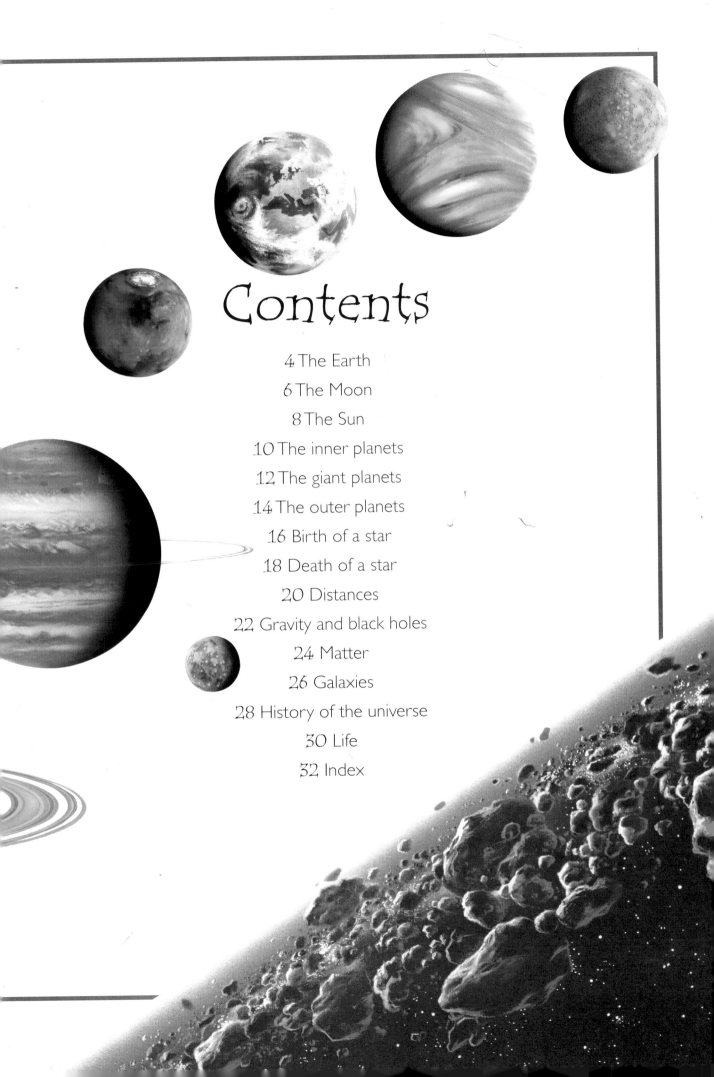

Contents

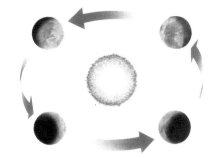

Earth's orbit round the Sun

As the Earth orbits the Sun, the hemisphere of the planet that faces the Sun has its summer. The hemisphere facing away is in winter.

How long is a day?
A day is the time Earth takes to turn once. The stars come back to the same place in the sky every 23 hours 56 minutes 4.09 seconds (the sidereal day). Our day (the solar day) is 24 hours because Earth is moving round the Sun, and must turn an extra 1° for the Sun to return to the same place in the sky.

Why does the Earth spin?
The Earth spins because it is falling around the Sun. As the Earth hurtles round the Sun, the Sun's gravity keeps it spinning, just as the Earth's gravity keeps a ball rolling downhill.

How did the Earth begin?

A ROUND FOUR AND A HALF BILLION YEARS AGO, NEITHER THE EARTH nor any of the other planets existed. There was just this vast dark very hot cloud of gas and dust swirling around the newly formed Sun. Gradually, the cloud cooled and the gas began to condense into billions of droplets. Slowly these droplets were pulled together into clumps by their own gravity – and they carried on clumping until all the planets, including the Earth, were formed. But it took another half a billion years before the Earth had cooled enough to form a solid crust with an atmosphere around it.

The early Earth was a fiery ball, then the surface cooled to form a hard crust.

Earth began life as hot gases and dust spiralling around the newborn Sun congealed into a ball.

How old is the Earth?
The Earth is about 4.6 billion years old. The oldest rock is about 3.8 billion years old. Scientists have also dated meteorites that have fallen from space, and must have formed at the same time as the Earth.

What is the Earth made of?

The Earth has a core of iron and nickel, and a rocky crust made mostly of oxygen and silicon. In between is the soft hot mantle of metal silicates, sulphides and oxides.

By 4 million years ago, the Earth's crust was covered in meteor craters and huge volcanoes.

The Earth cooled more, and the clouds of steam became water, creating vast oceans.

As the Earth cooled, it gave off gases and water vapour, which formed the atmosphere.

How big is the Earth?

Satellite measurements show it is 40,024 km (24,870 miles) around the equator and 12,578 km (7,927 miles) across. The diameter at the Poles is slightly less, by 43 km (26.7 miles).

What shape is the Earth?

The Earth is not quite a perfect sphere. Because it spins faster at the Equator than at the Poles, Earth bulges at the Equator. Scientists describe Earth's shape as 'geoid', which simply means Earth-shaped!

Who was Copernicus?

In the 1500s, most people thought the Earth was fixed in the centre of the universe, with the Sun and the stars revolving round it. Nicolaus Copernicus (1473-1543) was the Polish astronomer who first suggested the Earth was moving around the Sun.

Exactly how long is a year?

EVERY YEAR THE EARTH TRAVELS ONCE AROUND THE SUN. THIS EPIC JOURNEY measures 938,886,400 km (548,018,150 miles) and takes exactly 365.24 days, which gives us our calendar year of 365 days. To make up the extra 0.24 days, we add an extra day to our calendar at the end of February in every fourth year, which is called the leap year – and then we have to knock off a leap year every four centuries.

What's so special about the Earth?

The Earth is the only planet with temperatures at which water can exist on the surface and is the only planet with an atmosphere containing oxygen. Water and oxygen are both needed for life.

What is the Moon?

THE MOON IS THE EARTH'S NATURAL SATELLITE AND HAS CIRCLED AROUND IT for at least four billion years. It is a rocky ball about a quarter of Earth's size and is held in its orbit by mutual gravitational attraction. Scientists believe that the Moon formed when early in Earth's history a planet smashed into it. The impact was so tremendous that nothing was left of the planet but a few hot splashes thrown back up into space. Within a day of the smash, these splashes had been drawn together by gravity to form the Moon.

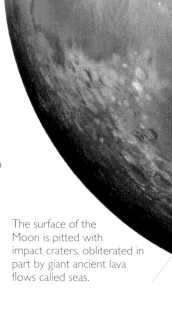

The Moon's surface is covered with a fine layer of dust.

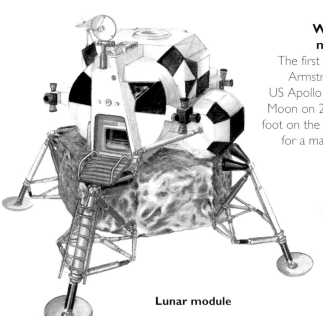

Who were the first men on the moon?
The first men on the Moon were Neil Armstrong and Buzz Aldrin of the US Apollo 11 mission who landed on the Moon on 21 July 1969. (As Armstrong set foot on the Moon, he said: 'This a small step for a man, a giant leap for mankind.')

The lunar module from the Apollo 15 mission was the astronauts' home during their brief stay on the Moon.

The surface of the Moon is pitted with impact craters, obliterated in part by giant ancient lava flows called seas.

Lunar module

What are the Moon's seas?
The large dark patches visible on the Moon's surface are called seas, but in fact they are not seas at all. They are huge plains formed by lava flowing from volcanoes that erupted early on in the Moon's history.

What is a lunar eclipse?
As the Moon goes round the Earth, sometimes it passes right into Earth's shadow, where sunlight is blocked off. This is a lunar eclipse. If you look at the Moon during this time, you can see the dark disc of the Earth's shadow creeping across the Moon.

How long is a month?
It takes the Moon 27.3 days to circle the Earth, but 29.53 days from one full moon to the next, because the Earth moves as well. A lunar month is the 29.53 days cycle. Calendar months are entirely artificial.

What is moonlight?
The Moon is by far the brightest thing in the night sky. But it has no light of its own. Moonlight is simply the Sun's light reflected off the white dust on the Moon's surface.

The Moon's phases

What's inside the Moon?

The Moon's mantle is now very cool compared to the Earth's.

The Moon's outer core is probably solid metal.

The Moon has an inner core of metal, very much smaller in relation to its size than Earth's.

Why does the Moon look like cheese?

The Moon looks like cheese because it is full of holes, and sometimes appears yellowish too. The holes are craters in the surface created when it was bombarded by huge rocks early on in its history.

The Moon has a crust of solid rock thicker than Earth's – up to 150 km (90 miles) thick on the side away from the Earth.

What is a harvest moon?

The harvest moon is the full moon nearest the autumnal equinox (when night and day are of equal length). This moon hangs bright above the eastern horizon for several evenings, providing a good light for harvesters.

Why does the sea have tides?

The Moon's gravity draws the oceans into an oval around the Earth, creating a bulge of water on each side of the world. These bulges stay beneath the Moon as the Earth spins round and so seem to run around the world, making the tide rise and fall as they pass.

What is a new moon?

THE MOON APPEARS TO CHANGE SHAPE DURING THE MONTH because, as it circles the Earth, we see its bright, sunny side from a different angle. At the new moon, the Moon is between Earth and the Sun, and we catch only a crescent-shaped glimpse of its bright side. Over the first two weeks of the month, we see more and more of the bright side (waxing) until full moon, when we see all its sunny side. Over the next two weeks, we see less and less (waning), until we get back to just a sliver – the old moon.

The phases of the Moon, from left to right: new moon, half moon (waxing), gibbous moon (waxing), full moon, gibbous moon (waning), half moon (waning), old moon.

What is the Sun?

T HE SUN IS AN AVERAGE STAR, JUST LIKE COUNTLESS OTHERS IN THE UNIVERSE.
It formed from gas left behind after an earlier, much larger star blew up and now, in middle-age, burns yellow and fairly steadily – giving the Earth daylight and remarkably constant temperatures. Besides heat and light, the Sun sends out deadly gamma rays, X-rays and ultraviolet, as well as infrared and radio waves. Fortunately we are shielded from these by Earth's magnetic field and atmosphere.

What is a solar eclipse?
A solar eclipse is when the Moon comes in between the Sun and the Earth, creating a shadow a few hundred kilometres wide on the Earth.

The photosphere is a sea of boiling gas. It gives the heat and light we experience on Earth.

How big is the Sun?
The Sun is a small to medium-sized star 1,392,000 km (0.86 million miles) in diameter. It weighs just under 2,000 trillion trillion tonnes.

What is the Sun's crown?
The Sun's crown is its corona, its glowing white hot atmosphere seen only as a halo when the rest of the sun's disc is blotted out by the Moon in a solar eclipse.

How hot is the Sun?
The surface of the Sun is a phenomenal 6,000°C (11,000°F), and would melt absolutely anything. But its core is thousands of times hotter at over 16 million°C (29 million°F)!

Sunspot

What makes the Sun burn?
The Sun gets its heat from nuclear fusion. Huge pressures deep inside the Sun force the nuclei (cores) of hydrogen atoms to fuse together to make helium atoms, releasing huge amounts of nuclear energy.

The chromosphere is a tenuous layer through which dart tongues called spicules, making it look like a flaming forest.

How old is the Sun?

The sun is a middle-aged star and probably formed about five billion years ago. It will probably burn for another five billion years and then die in a blaze so bright that the Earth will be scorched right out of existence.

What is the solar wind?

The solar wind is the stream of radioactive particles constantly blowing out from the Sun at hundreds of kilometres per second. (The Earth is protected from the solar wind by its magnetic field, but at the Poles the solar wind interacts with Earth's atmosphere to create the aurora borealis or northern lights.)

Beyond the chromosphere is the sun's ultra-thin halo of boiled-off gases called the corona.

Higher above the chromosphere are giant tongues of hot gases called prominences.

What are solar flares?

Flares are eruptions from the Sun's surface that fountain into space with the energy of one million atom bombs for about five minutes. (They are similar to solar prominences, the giant flame-like tongues of hot hydrogen that loop 100,000 km/60,000 miles into space.)

What are sunspots?

SUNSPOTS ARE DARK BLOTCHES SEEN ON THE SUN'S SURFACE. They are thousands of kilometres across, and usually occur in pairs. They are dark because they are slightly less hot than the rest of the surface. As the Sun rotates, they slowly cross its face – in about 37 days at the Equator and 26 days at the Poles. The average number of spots seems to reach a maximum every 11 years, and many scientists believe these sunspot maximums are linked to periods of stormier weather on Earth.

What's frightening about Mars's moons?

One night American astronomer Asaph Hall got fed up with studying Mars and decided to go to bed. But his domineering wife bullied him into staying up – and that night he discovered Mars's two moons. Mocking his fear of his wife, he named the moons Phobos (fear) and Deimos (panic).

Why is Mars red?

Mars is red because it is rusty. The surface contains a high proportion of iron dust, and this has been oxidized in the carbon dioxide atmosphere.

Mars

Mars is reddish with shadows visible here and there on the surface of the planet.

Is there life on Mars?

The Viking landers of the 1970s found not even the minutest trace of life. But in 1996, microscopic fossils of what might be mini-viruses were found in a rock from Mars. So who knows?

Three quarters of Earth's surface is covered in water, which is why it looks blue.

Earth

What canyon is bigger than the Grand Canyon?

A CANYON ON MARS! THE SURFACE OF MARS IS MORE STABLE THAN EARTH'S, and there is no rain or running water to wear down the landscape. So although it is only half the size of Earth, it has a volcano called Olympus Mons 24 km (15 miles) high – three times as high as Mount Everest. It also has a great chasm, discovered by the Mariner 9 space probe and called the Valles Marineris. This is over 4,000km (2,500 miles) long and four times as deep as America's Grand Canyon.

What is the air on Venus?

Venus's atmosphere would be deadly for humans. It is very deep, so the pressure on the ground is huge. It is made mainly of poisonous carbon dioxide and is also filled with clouds of sulphuric acid.

Why is Venus called the Evening Star?

Venus reflects sunlight so well it shines like a star. But because it is quite close to the Sun, we can only see it in the evening, just after the Sun sets. We can also see it just before sunrise.

What are the inner planets made of?

Each of the inner planets is formed a little bit like an egg – with a hard 'shell' or crust of rock, a 'white' or mantle of soft, semi-molten rock, and a 'yolk' or core of hot often molten iron and nickel.

Venus

Could you breathe on Mercury?

Not without your own oxygen supply. Mercury has almost no atmosphere – just a few wisps of sodium vapour – because gases are burned off by the nearby Sun.

How hot is Mercury?

Temperatures on Mercury veer from one extreme to the other because it has too thin an atmosphere to insulate it. In the day, temperatures soar to 430°C (800°F); at night they plunge to −180°C (−290°F).

Venus is a soft pinkish white ball with no features visible on the surface through its thick atmosphere.

Mercury

Mercury has virtually no atmosphere and its surface is pitted with craters like the Moon.

Sun

What are the inner planets?

THE INNER PLANETS ARE THE FOUR PLANETS IN THE SOLAR SYSTEM that are nearest to the Sun. These planets – Mercury, Venus, Earth and Mars – are small planets made of rock, unlike the bigger planets further out, which are made mostly of gas. Because they are made of rock, they have a hard surface a spaceship could land on, which is why they are sometimes called terrestrial (earth) planets. They all have a thin atmosphere, but each is very different.

What are the giant planets?

JUPITER AND SATURN – THE FIFTH AND SIXTH PLANETS OUT FROM THE SUN – are the giants of the solar system. Jupiter is twice as heavy as all the planets put together, and 1,300 times as big as the Earth. Saturn is almost as big. Unlike the inner planets, they are both made largely of gas, and only their very core is rocky. This does not mean they are vast cloud balls. The enormous pressure of gravity means the gas is squeezed until it becomes liquid and even solid.

How heavy is Saturn?
Saturn may be big, but because it is made largely of liquid hydrogen, it is also remarkably light, with a mass of 600 billion trillion tonnes. If you could find a big enough bath, it would float.

What is the Cassini division?
Saturn's rings occur in broad bands labelled with the letters A to G. In 1675, the astronomer Cassini spotted a dark gap between rings A and B, which is now called after him, the Cassini division.

When it rains on Saturn, it rains drops of liquid helium.

What are the giant planets made of?
Jupiter and Saturn are made largely of hydrogen and helium. On Jupiter, internal pressures are so great that most of the hydrogen is turned to metal.

Saturn

How windy is Saturn?
Saturn's winds are even faster than Jupiter's and roar round the planet at up to 1,800 kph (1,120 mph). But Neptune's are even faster!

Saturn's rings are made of billions of tiny chips of dust and ice.

The surface of one of Saturn's moons

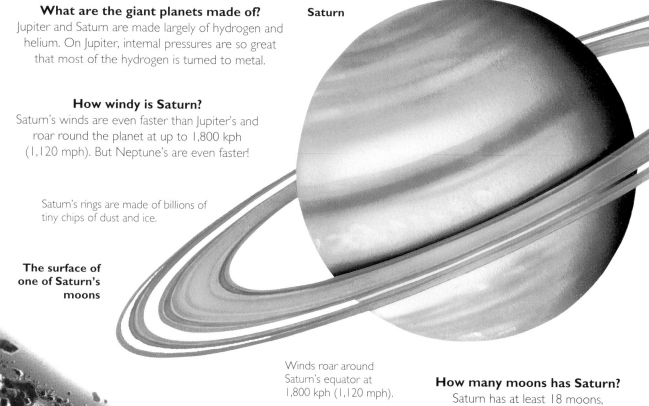

Winds roar around Saturn's equator at 1,800 kph (1,120 mph).

Saturn's moons are all blocks of ice, made dirty with dust and organic compounds, and the surface is barren.

How many moons has Saturn?
Saturn has at least 18 moons, including Iapetus, which is black on one side and white on the other, and Enceladus, which is covered in shiny beads of ice and shimmers like a cinema screen.

Why are astronomers excited about Titan?
Saturn's moon Titan is very special because it is the only moon in the solar system with an atmosphere.

One of Jupiter's moons passing in front of the planet.

High-speed winds whirl round Jupiter's surface, creating bands of cloud in the atmosphere.

Jupiter has a ring system, like Saturn, but much, much smaller.

Jupiter, the biggest planet in the solar system

Io, one of Jupiter's four Galilean moons

Volcanoes erupting sulphur on the surface of Io make it look like a pizza.

What are Saturn's rings?

SATURN'S RINGS ARE THE PLANET'S SHINING HALO, FIRST DISCOVERED IN 1659 by Dutch scientist Christiaan Huygens (1629-95). They are made of countless billions of tiny chips of ice and dust – few bigger than a refrigerator and most the size of ice cubes. The rings are incredibly thin – no more than than 50 m (164 ft) deep – yet they stretch way above Saturn's clouds, 7,000 km (4,350 miles) high, and over 74,000 km (46,000 miles) out into space. One of Saturn's rings is as thin as a piece of tissue paper being stretched over a football pitch.

How many moons has Jupiter?
Jupiter has 16 moons – four big ones, discovered by Galileo as long ago as 1610, and so called after him the Galilean moons, and 12 smaller ones.

What is Jupiter's red spot?
The Great Red Spot or GRS is a huge swirling storm in Jupiter's atmosphere 40,000 km (25,000 miles) across that has gone on in the same place for at least 330 years.

Could you land on Jupiter?
No. Even if your spaceship could withstand the enormous pressures, there is no surface to land on – the atmosphere merges unnoticeably into deep oceans of liquid hydrogen.

How fast does Jupiter spin?
Jupiter spins faster than any other planet. Despite its huge size, it turns right round in just 9.8 hours, which means the surface is moving at 45,000 kph (28,000 mph)!

How big is Jupiter?
Very big. Even though Jupiter is largely gas it weighs 320 times as much as the Earth and is 142,984 km (88,850 miles) in diameter.

What are the outer planets?

THE OUTER PLANETS ARE URANUS, NEPTUNE and Pluto, and Pluto's companion Charon. Unlike the other planets, these were completely unknown to ancient astronomers. They are so far away, and so faint, that Uranus was discovered only in 1781, Neptune in 1846, Pluto in 1930 and Charon as recently as 1978. Uranus and Neptune are gas giants like Jupiter and Saturn, but Pluto and Charon are rocky and were probably wandering asteroids trapped by the Sun's gravity within the outer reaches of the solar system.

Uranus

What's strange about Uranus?

Unlike any of the other planets, Uranus does not spin on a slight tilt. Instead it is tilted right over and rolls round the Sun on its side, like a giant bowling ball.

Uranus rolls on its side and is the seventh planet out from the Sun.

How long is a year on Neptune?

Neptune is so far from the Sun – over four billion km (2,800 million miles) at maximum – that its orbit takes 164,79 years. So Neptune's year is 164,79 of ours.

What's an asteroid?

Asteroids are the thousands of rocky lumps that circle round the Sun in a big band between Mars and Jupiter. The biggest, Ceres, is 1000 km (600 miles) across. Most are much smaller. Over 3,200 asteroids have been identified so far.

Why is Neptune green?

Neptune is greeny blue because its surface is completely covered in immensely deep oceans of liquid methane (natural gas).

At the heart of a comet is a nucleus of ice and dust often shaped like a lumpy potato, just a few km across.

What's a comet?

SPECTACULAR COMETS ARE JUST DIRTY ICEBALLS A FEW KILOMETRES ACROSS. Normally, they circle the outer reaches of the solar system. But occasionally, one of them is drawn in towards the Sun. As it hurtles towards the Sun, it melts and a vast tail of gas is blown behind it by the solar wind. We may see this spectacular tail in the night sky shining in the sunlight for a few weeks until it swings round the Sun and out of sight. The most recent comet was Hale-Bopp in 1997.

Neptune

Pluto and Charon

Neptune is so far from the Sun that surface temperatures drop to −210°C (−346°F).

How big is Pluto?
Pluto is very small, which is why it was so hard to spot. It is five times smaller than the Earth – just 2,284 km (1,419 miles) across – and 500 times lighter.

Who found Neptune?
Two mathematicians, John Couch Adams in England and Urbain le Verrier in France, predicted where Neptune should be from the way its gravity disturbed Uranus's orbit. Johan Galle in Berlin spotted it on 23 September, 1846.

A tail of ionized atoms is blown out millions of kms behind the comet by the solar wind.

A comet in the night sky

What is a meteorite?
Meteorites are lumps of rock from space big enough to penetrate the Earth's atmosphere and reach the ground without burning up.

Why do some stars throb?
The light from variable stars flares up and down. 'Cepheid' are big young stars that pulsate over a few days or a few weeks. 'RR Lyrae' variables are old yellow stars that vary over a few hours.

How are stars born?
Stars are born when clumps of gas in space are drawn together by their own gravity, and the middle of the clump is squeezed so hard that temperatures reach 10 million°C (18 million°F), so a nuclear fusion reaction starts.

How many stars are there?
It is hard to know how many stars there are in the universe, for the vast majority of them are much too far away to see. But astronomers guess there are about 200 billion billion.

Stars begin life when clumps of dust and gas in clouds called nebulae are pulled together by their own gravity.

Gravity squeezes these clumps. They begin to get hot.

Small clumps do not get very hot and soon fizzle out. If the clump is big enough, pressure in the centre boosts temperatures beyond 10 million°C (18 million°F).

Where are stars born?

STRETCHED THROUGHOUT SPACE ARE VAST CLOUDS OF DUST AND GAS CALLED NEBULAE. These clouds are 99 per cent hydrogen and helium with tiny amounts of other gases and minute quantities of icy, cosmic dust. Stars are born in the biggest of these nebulae, which are called giant molecular clouds. Here temperatures plunge to −263°C −441°F), just 10° short of absolute zero. These nebulae are thin and cold but contain all the materials needed to make a star.

What is the biggest star?
The biggest stars are the supergiants. Antares is 700 times as big as the Sun. There may be a star in the Epsilon system in the constellation of Auriga that is 3 billion km (1,860 million miles) across – 4,000 times as big as the Sun!

What are constellations?

Constellations are small patterns of stars in the sky, each with its own name. They have no real existence, but they help astronomers locate things in the night sky.

What makes stars glow?

Stars glow because the enormous pressure deep inside generates nuclear fusion reactions in which hydrogen atoms are fused together, releasing huge quantities of energy.

What is a star?

STARS ARE GIGANTIC GLOWING BALLS OF GAS, SCATTERED THROUGHOUT SPACE. They burn for anything from a few million to tens of billions of years. The nearest star, apart from the Sun, is over 40 trillion km (25 trillion miles) away. They are all so distant that we can see stars only as pinpoints of light in the night sky – even through the most powerful telescope. As far as we can see there are no other large objects in the universe.

Nuclear fusion begins as hydrogen atoms fuse together to make helium. The heat from the fusion makes the star shine.

In medium-sized stars, like our Sun, the heat generated in the core pushes gas out as hard as gravity pulls it in, so the star stabilizes and burns steadily for billions of years.

After 10 billion years or so, all the hydrogen in the star's core is burned up, and the core shrinks as it begins to burn helium.

How hot is a star?

The surface temperature of the coolest stars is below 3,500°C (6,300°F); that of the hottest, brightest stars is over 40,000°C (72,000°F).

What makes stars twinkle?

Stars twinkle because the Earth's atmosphere is never still, and starlight twinkles as the air wavers. Light from the nearby planets is not distorted as much, so they don't twinkle.

What colour are stars?

It depends how hot they are. The colour of medium-sized stars varies along a band on a graph called the main sequence – from hot and bright blue-white stars to cool and dim red stars.

What happens when stars die?

STARS DIE WHEN THEY HAVE EXHAUSTED THEIR VAST SUPPLIES OF NUCLEAR FUEL. When the hydrogen runs out, they switch to helium. When the helium runs out, they quickly exhaust any remaining nuclear energy and either blow up, shrink or go cold. Just how long it takes to reach this point depends on the star. The biggest stars have masses of nuclear fuel but live fast and die young. The smallest stars have little nuclear fuel but live slow and long. A star twice as big as the Sun lives a tenth as long. The biggest stars live just a few million years.

What is a pulsar?
Pulsars are stars that flash out intense radio pulses every ten seconds or less as they spin rapidly round. They are thought to be very, very dense dying stars called neutron stars.

How old are the stars?
Stars are dying and being born all the time. Big bright stars live only ten million years. Medium-sized stars like our Sun live ten billion years.

What is a red giant?
Red giants are the huge, cool stars formed as surface gas on medium-sized stars nearing the end of their life swells up.

The outer layers cool off and swell so that the star grows into a cool red giant star.

The biggest stars go on swelling into supergiants. Pressure at the centre becomes so immense that carbon and silicon fuse to make iron.

Once iron forms in its centre, the star does not release energy but absorbs it – the star suddenly and catastrophically collapses in seconds.

What are the oldest stars?
The oldest stars we know of are not stars at all, but simply look like them because they are very bright but very far away. These are 'quasi-stellar radio objects' or quasars. Some are so far away the light we see left them 13 billion years ago.

The remains of a supernova collapses to become a pulsar, a star made mostly of atomic nuclear particles called neutrons.

The collapse of a supergiant triggers an explosion like a huge nuclear bomb, called a supernova.

The pulsar spins rapidly, beaming out pulses of radiation like a lighthouse.

What is a white dwarf?
White dwarfs are the small white stars formed as stars smaller than our Sun lose their surface gas altogether and shrink.

For a few weeks, the exploding supernova shines as brightly as billions of suns.

What is a supernova?

A SUPERNOVA IS A GIGANTIC EXPLOSION. IT FINISHES OFF A SUPERGIANT STAR. For just a brief moment, the supernova flashes out with the brilliance of billions of suns. Supernovae are rare, and usually visible only through a telescope. But in 1987, for the first time in 400 years, one called Supernova 1987A was visible with the naked eye for nine months.

What are neutron stars?
Neutron stars are all that remains of a supergiant star after a supernova explosion. They are tiny unimaginably dense stars that often become pulsars.

What is a light-year?

A light-year is 9,460,000,000,000 km. This is the distance light can travel in a year, at its constant rate of 300,000 km per second.

What is the furthest star we can see?

The furthest objects we can see in space are quasars, which may be over 13 billion light-years away.

What is a parsec?

A parsec is 3.26 light-years. Parsecs are parallax distances – distances worked out geometrically from slight shifts of a star's apparent position as the Earth moves round the Sun.

Distances within the solar system can be given in kilometres or miles.

Distances to nearby stars are measured in light-years.

What is red shift?

When a galaxy is moving rapidly away from us the waves of light become stretched out – that is, they become redder. The greater this red shift, the faster the galaxy is moving away from us.

Distances to the furthest galaxies are measured in billions of light-years.

Are the stars getting further away?

Analysis of red shifts has shown us that every single galaxy is moving away from us. The further away the galaxy, the faster it is moving away from us. The most distant galaxies are receding at almost the speed of light.

How far away is the Moon?

AT ITS NEAREST, THE MOON IS 356,410 KM (221,463 MILES) AWAY FROM EARTH; at its furthest, it is 406,697 km (252,710 miles) away. This is measured accurately by a laser beam bounced off mirrors left on the Moon's surface by Apollo astronauts and Soviet lunar probes. The distance is shown by how long it takes the beam to travel to the Moon and back.

Pluto and Charon

Neptune

Saturn

Uranus

Pluto and its companion Charon (top) are 5,914 million km from the Sun.

Saturn (above) is 1,427 million km from the Sun.
Uranus (left) is 2,871 million km from the Sun.

How did astronomers first estimate the Sun's distance?

Sun

Mercury

Mercury is 57.9 million km from the Sun.

Venus is 108.2 million km from the Sun.

Venus

IN 1672, TWO ASTRONOMERS, CASSINI IN FRANCE AND RICHER IN GUIANA, noted the exact position of Mars in the skies. They could work out how far away Mars is from the slight difference between their two measurements. Once they knew this, they could work out by simple geometry the distance from Earth to the Sun. Cassini's estimate was a mere 7 per cent too low.

How far away is the Sun?
The distance varies between 147 and 152 million km (91 to 94 million miles) from Earth. This is measured very accurately by bouncing radar waves off the planets.

How far is it to the nearest star?
The nearest star is Proxima Centauri, which is 4.3 light-years away – 40 trillion km.

Mars

Earth

Jupiter

The dome rotates, so the telescope can track stars across the sky.

Observatory

Jupiter is 778.3 million km from the Sun.

How do astronomers measure distance?
For nearby stars, they use parallax (see What is a parsec?). With middle distance stars, they look for standard candles, stars whose brightness they know. The dimmer it looks, compared to how bright it should look, the further away it is.

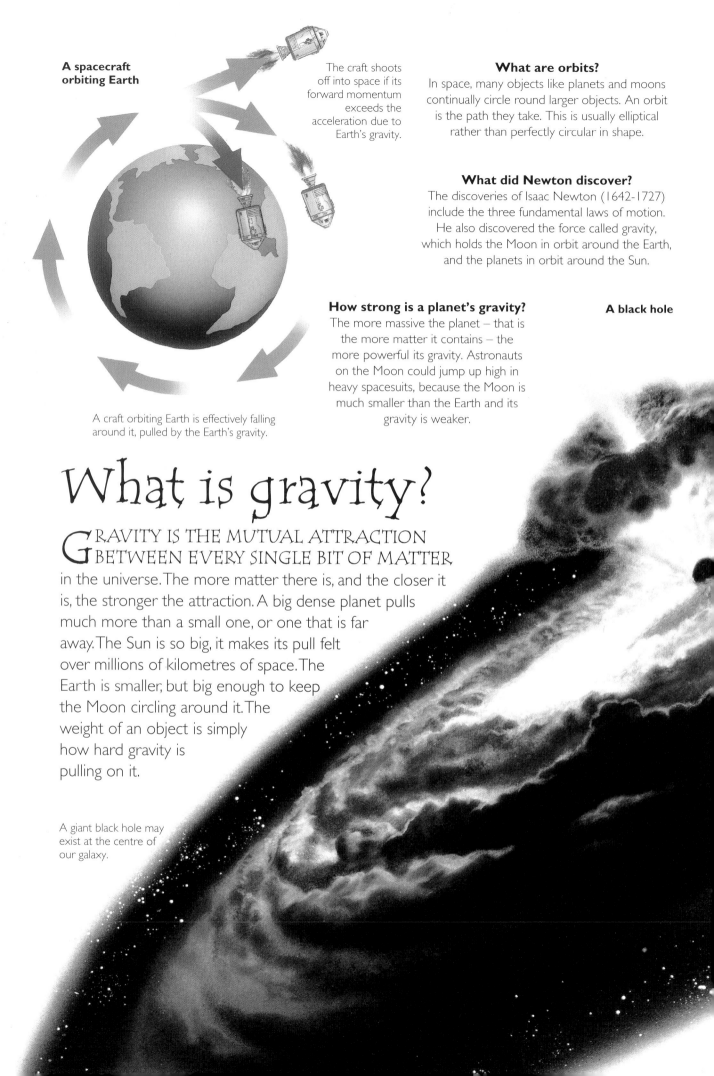

A spacecraft orbiting Earth

The craft shoots off into space if its forward momentum exceeds the acceleration due to Earth's gravity.

A craft orbiting Earth is effectively falling around it, pulled by the Earth's gravity.

What are orbits?

In space, many objects like planets and moons continually circle round larger objects. An orbit is the path they take. This is usually elliptical rather than perfectly circular in shape.

What did Newton discover?

The discoveries of Isaac Newton (1642-1727) include the three fundamental laws of motion. He also discovered the force called gravity, which holds the Moon in orbit around the Earth, and the planets in orbit around the Sun.

How strong is a planet's gravity?

The more massive the planet – that is the more matter it contains – the more powerful its gravity. Astronauts on the Moon could jump up high in heavy spacesuits, because the Moon is much smaller than the Earth and its gravity is weaker.

A black hole

What is gravity?

GRAVITY IS THE MUTUAL ATTRACTION BETWEEN EVERY SINGLE BIT OF MATTER in the universe. The more matter there is, and the closer it is, the stronger the attraction. A big dense planet pulls much more than a small one, or one that is far away. The Sun is so big, it makes its pull felt over millions of kilometres of space. The Earth is smaller, but big enough to keep the Moon circling around it. The weight of an object is simply how hard gravity is pulling on it.

A giant black hole may exist at the centre of our galaxy.

What happens inside a black hole?

Nothing that goes into a black hole comes out, and there is a point of no return called the event horizon. If you went beyond this you would be 'spaghettified' – stretched long and thin until you were torn apart by the immense gravity.

How big is a black hole?

The singularity at the heart of a black hole is infinitely small. The size of the hole around it depends on how much matter went into forming it. The black hole at the heart of our galaxy may be around the size of the solar system.

How many black holes are there?

No one really knows. Because they trap light they are hard to see. But there may be one at the heart of every galaxy.

The black hole contains so much matter in such a small space that its gravitational pull even drags in light.

We may be able to spot a black hole from the powerful radio signals emitted by stars being ripped to shreds as they are sucked in.

What is a black hole?

IF A SMALL STAR IS VERY DENSE, IT MAY BEGIN TO SHRINK UNDER THE PULL of its own gravity. As it shrinks, it becomes denser and denser and its gravity becomes more and more powerful – until it shrinks to a single tiny point of infinite density called singularity. The gravitational pull of a singularity is so immense that it pulls space into a 'hole' like a funnel. This is the black hole, which sucks in everything that comes near it with its huge gravitational force – including light, which is why it is a 'black' hole.

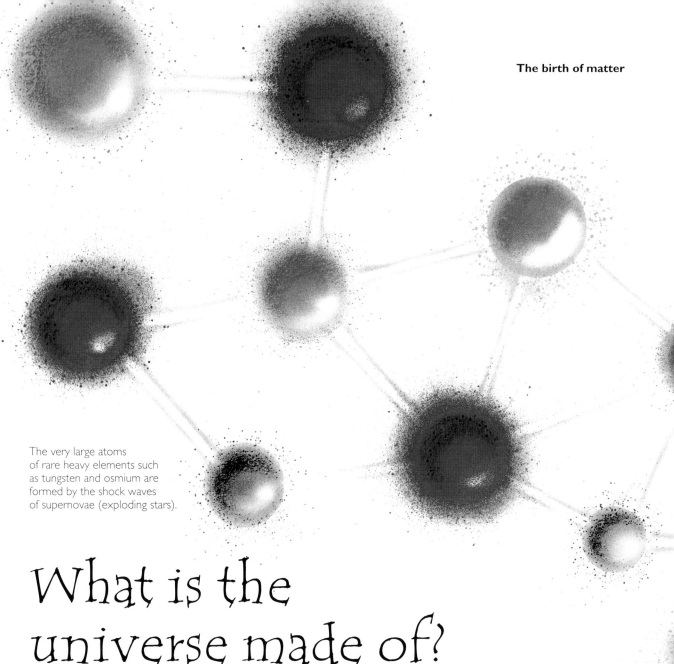

The very large atoms of rare heavy elements such as tungsten and osmium are formed by the shock waves of supernovae (exploding stars).

What is the universe made of?

THE STARS AND CLOUDS ARE MADE ALMOST 100 PER CENT OF HYDROGEN AND HELIUM, the lightest and simplest elements of all. All the other elements are relatively rare. But some, such as carbon, oxygen, silicon, nitrogen and iron can form important concentrations. This happens in the few rocky planets like Earth, where iron, oxygen and magnesium are the most common elements. Carbon, a scarcer element, is the one on which all life forms are based.

Larger atoms such as beryllium, carbon and oxygen are made when nuclear reactions inside stars force the nuclei of helium atoms together.

What are particles?
Particles are the tiny concentrations of energy from which all matter is made up, of which atoms are the largest. There are hundreds of kinds of particles, but all but atoms are too small to see, even with the most powerful microscope.

How were atoms made?
Atoms of hydrogen and helium were made in the early days of the universe when quarks in the matter soup joined together. All other atoms were made as atoms were fused together by the intense heat and pressure inside stars.

How was iron made?
Iron was forged in the heart of supergiant stars near the end of their life when the immense pressures there force carbon atoms together.

24

What holds everything together?

EVERYTHING IN THE UNIVERSE IS HELD TOGETHER BY FOUR invisible forces. Two of them – gravity and electromagnetism – are familiar in everyday life. The other two – the strong and weak nuclear forces – are unfamiliar because they operate only inside the invisibly small nucleus of the atom, holding it all together.

What was the first element?
The first element to form was hydrogen, which has the simplest and lightest atom of all. It formed within three minutes of the dawn of the universe.

What's the smallest particle?
The smallest particle inside the nucleus is the quark. It is less than 10^{-20} m across – which means a line of ten billion billion of them would be less than a metre long.

What is anti-matter?
Anti-matter is the mirror image of ordinary matter. If matter and anti-matter meet, they annihilate each other. Fortunately, there is very little anti-matter around.

First to form at the birth of the universe were countless particles much smaller than atoms, such as quarks.

Some of these particles fused together to form the first atoms, hydrogen and helium.

In the first few moments of the universe's existence, there was no matter – only seething, incredibly hot space.

Spiral galaxy

Spiral galaxies are spinning Catherine wheel spirals like our Milky Way.

Elliptical galaxy

Elliptical galaxies are shaped like rugby balls and are the oldest galaxies of all.

Barred spiral galaxy

Arms trail from this type of galaxy like water from a spinning garden sprinkler.

What is a galaxy?

OUR SUN IS JUST ONE OF A MASSIVE COLLECTION OF TWO BILLION STARS arranged in a shape like a fried egg, 100,000 light-years across. This collection is called the Galaxy because we see it in the band of stars across the night sky called the Milky Way. (Galaxy comes from the Greek for milky.) But earlier this century it was realized that the Galaxy is just one of millions of similar giant star groups scattered throughout space, which we also call galaxies. The nearest is the Andromeda galaxy.

What is a spiral galaxy?
A spiral galaxy is a galaxy that has spiralling arms of stars like a gigantic Catherine wheel. They trail because the galaxy is rotating. Our Galaxy is a spiral galaxy.

The Milky Way

What is the Milky Way?
The Milky Way is a pale blotchy white band that stretches right across the night sky. A powerful telescope shows it is made of thousands of stars, and is actually an edge-on view of our Galaxy.

Where is the Earth?
The Earth is just over half way out along one of the spiral arms of the Galaxy, about 30,000 light-years from the centre.

What are star clusters?
Stars are rarely entirely alone within a galaxy. Most are concentrated in groups called clusters. Globular clusters are big and round. Galactic clusters are small and formless.

If we could see the Milky Way from above, we would see that it is a giant spiral galaxy.

Irregular galaxy

Irregular galaxies are galaxies that have no particular shape at all.

What are double stars?
Our Sun is alone in space, but many stars have one or more nearby companions. Double stars are called binaries.

What exactly are nebulae?

NEBULAE ARE GIANT CLOUDS OF GAS AND DUST SPREAD THROUGHOUT the galaxies. Some of them we see through telescopes because they shine faintly as they reflect starlight. With others, called dark nebulae, we see only inky black patches hiding the stars behind. This is where stars are born. A few – called glowing nebulae – glow faintly of their own accord as the gas within them is heated by nearby stars.

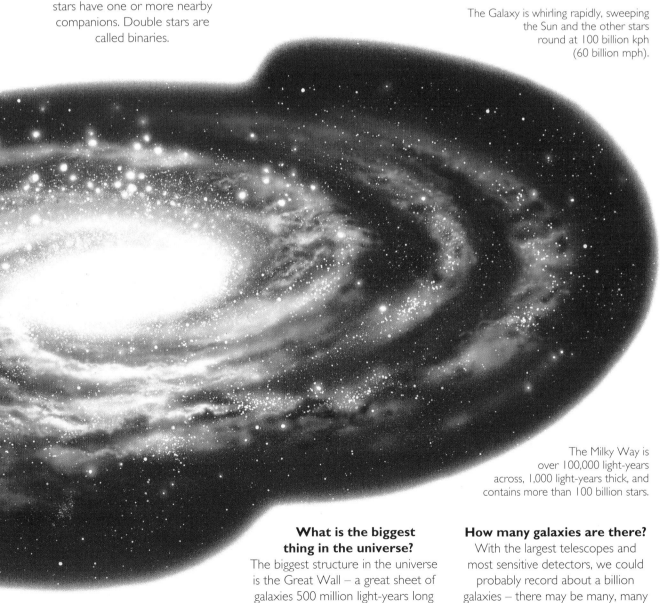

The Galaxy is whirling rapidly, sweeping the Sun and the other stars round at 100 billion kph (60 billion mph).

The Milky Way is over 100,000 light-years across, 1,000 light-years thick, and contains more than 100 billion stars.

What is the biggest thing in the universe?
The biggest structure in the universe is the Great Wall – a great sheet of galaxies 500 million light-years long and 16 million light-years thick.

How many galaxies are there?
With the largest telescopes and most sensitive detectors, we could probably record about a billion galaxies – there may be many, many more beyond their limits.

How do we know what it was like?

We know partly by mathematical calculations, and partly by experiments in huge machines called colliders and particle accelerators. These recreate conditions in the early universe by using magnets to accelerate particles to astonishing speeds in a tunnel, and then crash them together.

In the beginning there was a ball smaller than an atom. It grew as big as a football as it cooled from infinity to ten billion billion billion °C

How long will the universe last?

It depends how much matter it contains. If there is more than the 'critical density', gravity will put a brake on its expansion, and it may soon begin to contract again to end in a Big Crunch. If there is much less, it may go on expanding forever.

What was there before the universe?

No one has a clue. Some people think there was an unimaginable ocean beyond space and time of potential universes continually bursting into life, or failing. Ours succeeded.

What is inflation?

Inflation was a period in the first few trillionths of a trillionth of a trillionth of a second in the life of the universe, when space swelled up enormously, before there was matter and energy to fill it.

Can we see the Big Bang?

Astronomers can see the galaxies hurtling away in all directions. They can also see the afterglow – low level microwave radiation coming at us from all over the sky, called the microwave background.

After a split second inflation began as space swelled a thousand billion billion billion times in less than a second – from the size of a football to something bigger than a galaxy.

What was the universe like at the beginning?

THE EARLY UNIVERSE WAS VERY SMALL, BUT IT CONTAINED ALL THE MATTER and energy in the universe today. It was a dense and chaotic soup of tiny particles and forces – and instead of the four forces scientists know today, there was just one superforce. But this original universe lasted only a split second. After just three trillionths of a trillionth of a trillionth of a second, the superforce split up into separate forces.

What was the Big Bang?

In the beginning, all the universe was squeezed into an unimaginably small, hot, dense ball. The Big Bang was when this suddenly began to swell explosively, allowing first energy and matter, then atoms, gas clouds and galaxies to form. The universe has been swelling ever since.

Why is the universe getting bigger?

We can tell the universe is getting bigger because every galaxy is speeding away from us. Yet the galaxies themselves are not moving – the space in between them is stretching.

How old is the universe?

How did the first galaxies and stars form?
They formed from curdled lumps of clouds of hydrogen and helium – either as clumps broke up into smaller more concentrated clumps, or as concentrations within the clumps drew together.

WE KNOW THAT THE UNIVERSE IS GETTING BIGGER AT A CERTAIN RATE by observing how fast distant galaxies are moving. By working out how long it took everything to expand to where it is now, we can wind the clock back to the time when the universe was very, very small indeed. This suggests that the universe is between 13 and 15 billion years old. However, studies of globular clusters suggest some stars in our galaxy may be up to 18 billion years old.

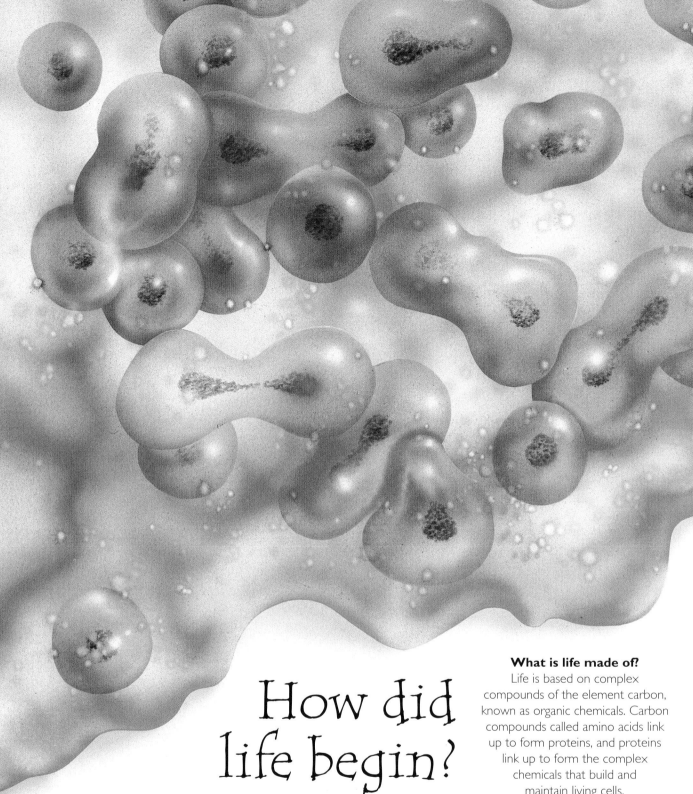

How did life begin?

SCIENTIFIC EXPERIMENTS IN THE 1950s SHOWED HOW LIGHTNING FLASHES might create amino acids, the basic chemicals of life, from the waters and gases of the early Earth. But no one knows how these chemicals joined up to become 'self-replicating' – that is, able to make copies of themselves. This is the key to life – which remains a mystery. However, the first lifeforms were probably tiny bacteria called Archebacteria, which thrive in very hot, chemically rich places.

What is life made of?
Life is based on complex compounds of the element carbon, known as organic chemicals. Carbon compounds called amino acids link up to form proteins, and proteins link up to form the complex chemicals that build and maintain living cells.

Where did the materials of life come from?
It used to be thought that organic chemicals all originated on Earth, but traces of all kinds of organic chemicals have been detected in giant molecular clouds, including formaldehyde, alcohol, and also acetaldehyde, one of the components of amino acids.

Where did life come from?

Most scientists think life on Earth began on Earth – in the oceans or in volcanic pools. But some think the Earth was seeded by micro-organisms from space.

What is SETI?

SETI is the Search for Extra-Terrestrial Intelligence project – designed to continually scan radio signals from space and pick up any signs of intelligence – signals that have a pattern, but are not completely regular like those from pulsating stars.

Why is the universe like it is?

The amazing chance that life exists on Earth has made some scientists wonder if only a universe like ours could contain intelligent life. This is called the weak anthropic principle. Some go further and say that the universe is constructed in such a way that intelligent life must develop at some stage. This is called the strong anthropic principle.

The first lifeforms

The first lifeforms were little more than simple chemical molecules surrounded by a membrane.

What is DNA?

Dioxyribonucleic acid, the most remarkable chemical in the universe – the tiny molecule on which all life is based. It is shaped a bit like a very long rope ladder, with two strands twisted together in a spiral, linked by 'rungs' of four different chemical bases. The order of these bases is a chemical code that provides all the instructions needed for life.

Is there life on other planets?

ORGANIC CHEMICALS ARE WIDESPREAD, AND THE CHANCES ARE THAT IN SUCH a large universe there are many planets, like Earth, suitable for nurturing life. But no one knows if life arose on Earth by a fantastic and unique chain of chance events – or whether it is fairly likely to happen given the right conditions. If the shapes found in Martian rock in 1996 really do prove to be lifeforms, the universe is probably teeming with life.

Are there any other planets like Earth?

There is no other planet like Earth anywhere in the solar system. Recently, though, planets have been detected circling other stars nearby. But they are much too far away for us to know anything about them at all.

How are we looking for extra-terrestrial life?

Since possible fossils of microscopic life were found in Martian rock in 1996, scientists have hunted for other signs of organisms in rocks from space. Future probes to Mars are also being designed to drill into the Martian surface and look for signs of microscopic life below ground.

What does an alien look like?

At the moment, the only aliens we are likely to encounter are very, very small and look like viruses.

Two-mouthed creature from outer space

No one knows what creatures from elsewhere in the universe would look like – but the chances are they would look pretty strange.

Index